I0516534

Tucholsky Wagner Zola Scott Sydow Freud Schlegel
Turgenev Wallace Fonatne
Twain Walther von der Vogelweide Fouqué Friedrich II. von Preußen
Weber Freiligrath Frey
Fechner Kant Ernst Frommel
Fichte Weiße Rose von Fallersleben Richthofen
Hölderlin
Engels Fielding Eichendorff Tacitus Dumas
Fehrs Faber Flaubert Ebner Eschenbach
Eliasberg
Feuerbach Maximilian I. von Habsburg Fock Eliot Zweig Vergil
Ewald
Goethe Elisabeth von Österreich London
Mendelssohn Balzac Shakespeare Dostojewski Ganghofer
Lichtenberg Rathenau Doyle Gjellerup
Trackl Stevenson Hambruch
Mommsen Tolstoi Lenz Droste-Hülshoff
Thoma Hanrieder
Dach Verne von Arnim Hägele Hauff Humboldt
Karrillon Reuter Rousseau Hagen Hauptmann Gautier
Garschin
Damaschke Defoe Hebbel Baudelaire
Descartes Hegel Kussmaul Herder
Wolfram von Eschenbach Dickens Schopenhauer Rilke George
Darwin Melville Grimm Jerome Bebel
Bronner Campe Horváth Aristoteles Proust
Bismarck Vigny Barlach Voltaire Federer Herodot
Gengenbach Heine
Storm Casanova Tersteegen Grillparzer Georgy
Chamberlain Lessing Gilm
Brentano Langbein Gryphius
Strachwitz Claudius Schiller Lafontaine Kralik Iffland Sokrates
Bellamy Schilling
Katharina II. von Rußland Gerstäcker Raabe Gibbon Tschechow
Löns Hesse Hoffmann Gogol Wilde Gleim Vulpius
Luther Heym Hofmannsthal Klee Hölty Morgenstern Goedicke
Roth Heyse Klopstock Homer Kleist
Luxemburg Puschkin Mörike
La Roche Horaz Musil
Machiavelli Kierkegaard Kraft Kraus
Navarra Aurel Musset Lamprecht Kind Kirchhoff Hugo Moltke
Nestroy Marie de France Laotse Ipsen Liebknecht
Nietzsche Nansen Ringelnatz
Marx Lassalle Gorki Klett Leibniz
von Ossietzky May vom Stein Lawrence Irving
Petalozzi Platon Knigge
Pückler Michelangelo Kock Kafka
Sachs Poe Liebermann
de Sade Praetorius Mistral Zetkin Korolenko

The publishing house tredition has created the series **TREDITION CLASSICS**. It contains classical literature works from over two thousand years. Most of these titles have been out of print and off the bookstore shelves for decades.

The book series is intended to preserve the cultural legacy and to promote the timeless works of classical literature. As a reader of a **TREDITION CLASSICS** book, the reader supports the mission to save many of the amazing works of world literature from oblivion.

The symbol of **TREDITION CLASSICS** is Johannes Gutenberg (1400 – 1468), the inventor of movable type printing.

With the series, tredition intends to make thousands of international literature classics available in printed format again – worldwide.

All books are available at book retailers worldwide in paperback and in hardcover. For more information please visit: www.tredition.com

tredition was established in 2006 by Sandra Latusseck and Soenke Schulz. Based in Hamburg, Germany, tredition offers publishing solutions to authors and publishing houses, combined with worldwide distribution of printed and digital book content. tredition is uniquely positioned to enable authors and publishing houses to create books on their own terms and without conventional manufacturing risks.

For more information please visit: www.tredition.com

A Course In Wood Turning

Archie S. Milton

Imprint

This book is part of the TREDITION CLASSICS series.

Author: Archie S. Milton
Cover design: toepferschumann, Berlin (Germany)

Publisher: tredition GmbH, Hamburg (Germany)
ISBN: 978-3-8491-8518-3

www.tredition.com
www.tredition.de

Copyright:
The content of this book is sourced from the public domain.

The intention of the TREDITION CLASSICS series is to make world literature in the public domain available in printed format. Literary enthusiasts and organizations worldwide have scanned and digitally edited the original texts. tredition has subsequently formatted and redesigned the content into a modern reading layout. Therefore, we cannot guarantee the exact reproduction of the original format of a particular historic edition. Please also note that no modifications have been made to the spelling, therefore it may differ from the orthography used today.

A COURSE IN WOOD TURNING

By
ARCHIE S. MILTON

OTTO K. WOHLERS

PREFACE

This book is the outgrowth of problems given to high school pupils by the writers, and has been compiled in logical sequence. Stress is laid upon the proper use of tools, and the problems are presented in such a way that each exercise, or project, depends somewhat on the one preceding. It is not the idea of the writers that all problems shown should be made, but that the instructor select only such as will give the pupils enough preliminary work in the use of the tools to prepare them for other models following.

The related matter on the care of the lathe and tools, the grinding of chisels, the polishing of projects, and the specific directions and cautions for working out the various exercises and projects with the drawings, make the book not only valuable for reference, but also as a class text to be studied in connection with the making of projects. The drawings show exact dimensions and are tabulated in the upper right-hand corner in such a way that they may be used in a filing case if desired. At least two designs are shown for each model, and these may be used as suggestions from which students, with the aid of the instructor, may work out their own designs.

The book has been divided into two parts: (A) Spindle Turning, and (B) Face-Plate Turning. The same order is followed in each part; the related information is supplied where required as the pupil progresses.

Part A takes up the following: (I) Exercises; (II) Models, involving the same tool processes, only in a somewhat different degree; (III) Oval Turning, explaining the use of two centers; (IV) Duplicate Turning, where identical pieces are turned.

Part B is arranged as follows: (I) Exercises; (II) Models, which are an application of cuts in exercises that involve only face-plate work; (III) Models, which require chucking; (IV) Assembling Exercises, involving spindle turning, face-plate work and chucking; (V) Spiral Turning, showing the method of turning a spiral on the lathe.

The ultimate aim of this book is to give, through the exercises and problems, a thorough understanding of the principles of wood turning by gradually developing the confidence of the pupil in the complete control of his tools, at the same time suggesting harmonious lines in design which will lead to other ideas in designing problems.

TABLE OF CONTENTS

CHAPTER I
Introductory
--Commercial and Educational Values of Wood Turning
--Elements of Success

CHAPTER II
The Lathe
--Care of the Lathe
--Speed of the Lathe
--Method of Figuring the Diameter of Pulleys
--Rules for Finding the Speeds and Sizes of Pulleys
--Points on Setting Up the Lathe and Shafting

CHAPTER III
Wood Turning Tools
--Grinding and Whetting Turning Tools
--The Gouge
--The Parting Tool
--Scraping Tools

CHAPTER IV
Spindle Turning
--Centering Stock
--Clamping Stock in the Lathe
--Adjusting the Tool Rest
--Position of the Operator at the Lathe
--Holding the Tools
--Use of the Tools in Spindle Turning

CHAPTER V
Tool Processes in Spindle Turning
--The Roughing Cut
--The Sizing Cut
--The Smoothing Cut
--Testing for Smoothness
--Measuring for Length
--Squaring Ends
--Cutting Off

--Shoulder Cuts
--Taper Cuts
--V Cuts-Concave Cuts
--Convex Cuts
--Combination Cuts
--Chisel Handles
--Mallets and Handles
--Vise Handles

CHAPTER VI
Oval Turning
--Tool Operations

CHAPTER VII
Duplicate Turning
--Use of Measuring Stick
--Use of Templets

CHAPTER VIII
Finishing and Polishing
--Ordinary Cabinet Finishing
--French Polishing
--Method of Applying French Polish

CHAPTER IX
Face-Plate and Chuck Turning
--Methods of Fastening Stock
--Small Single Screw Face-Plate
--Large Surface Screw Face-Plate
--Gluing to Waste Stock
--Lathe Adjustments
--Position of Tool Rest

CHAPTER X
Tool Processes in Face-Plate and Chuck Turning
--Straight Cuts
--Roughing Off Corners
--Calipering for Diameter
--Smoothing Cut
--Roughing Cut on the Face
--Smoothing the Face
--Laying Off Measurements

--External Shoulders
--Internal Shoulders
--Taper Cuts
--V Cuts
--Concave Cuts
--Convex Cuts
--Combination Cuts
--Use of Scraping Tools
--Internal Boring
--Turning a Sphere

CHAPTER XI

Spiral Turning
--Single Spiral, Straight Shaft
--Tapered Shaft
--Double Spiral, Tapered Shaft
--Double Spiral, Straight Shaft
--Double Groove Spiral, Straight Shaft

PLATES--SPINDLE TURNING.

Straight Cuts
Shoulder Cuts
Taper Cuts
V Cuts
Concave Cuts
Convex Cuts
Combination Cuts
Chisel Handles
Cabinet File Handle
Scratch Awl Handle
Carving Tool Handle
Turning Chisel Handle
Mallets
Gavels
Darning Eggs
Stocking Darner
Potato Masher
Rolling Pins
Vise Handle
Screw Driver Handles
Pene Hammer Handle
Claw Hammer Handle
Indian Clubs
Dumb Bells
Ten Pins
Drawer Pulls

PLATES--CHUCK TURNING.

Straight Cuts

Shoulder Cuts

Taper Cuts

V Cuts

Concave Cuts

Convex Cuts

Combination Cuts

Match Boxes

Pin Trays

Hair Pin Receivers

Hat Pin Receivers

Ornamental Vases

Spinnet

Towel Rings

Card Trays

Picture Frames

Nut Bowls

Napkin Rings

Jewel Boxes

Collar Boxes

Sphere

Checker Men

Candle Sticks

Shaving Stands

Reading Lamp Stands

Pedestal

Smokers' Stands

Pin Cushion and Spoon Holder

Chess Men

Pedestals
Electric Reading Lamps
Magazine Holders

CLASSIFICATION OF PLATES

A. SPINDLE TURNING
I. Exercises
1. Straight Cuts, a
2. Shoulder Cuts, a-b-c-d
3. Taper Cuts, a-b-c-d-e-f
4. V Cuts, a-b
5. Concave Cuts, a-b-c
6. Convex Cuts, a-b-c-d
7. Combination Cuts, a-b-c
II. Models
1. Chisel Handles, a-b-c
Cabinet File Handle, d
Scratch Awl Handle, e
Carving Tool Handle, f
Turning Chisel Handle, g
2. Mallets, a-b
3. Gavels, a-b-c-d
4. Stocking Darners, a-b
Darning Egg, c
5. Potato Mashers, a-b
6. Rolling Pins, a-b
7. Vise Handles, a
III. Oval Turning
1. Screw-driver Handles, a-b
2. Hammer Handles
Penne Hammer Handle, a
Claw Hammer Handle, b
IV. Duplicate Turning
1. Indian Clubs, a-b
2. Dumb-bells, a-b
3. Tenpins, a
4. Drawer Pulls, a-b
B. FACE-PLATE AND CHUCK TURNING
I. Exercises
1. Straight Cuts, a-b
2. Shoulder Cuts, a-b

3. Taper Cuts, a-b
4. V Cuts, a-b
5. Concave Cuts, a-b
6. Convex Cuts, a-b
7. Combination Cuts, a-b-c
II. Face-Plate Models
1. Match Boxes, a-b-c
2. Pin Trays, a-b
3. Hair Pin Receivers, a-b
4. Hat Pin Receivers, a-b
5. Ornamental Vases, a-b-c
6. Spinnet, (game) a
III. Chuck Models
1. Towel Rings, a-b-c
2. Card Trays, a-b-c-d
3. Picture Frames, a-b-c-d
4. Nut Bowels, a-b-c-d
5. Napkin Rings, a-b-c
6. Jewel Boxes, a-b-c-d-e-f-g-h
7. Collar Boxes, a-b-c
8. Spheres, a
9. Checker Men, a
IV. Assembling Exercises
1. Candle Sticks, a-b-c-d-e
2. Shaving Stands, a-a'-b-b'
3. Reading Lamp Stands, a-b-c
4. Pedestals, a
5. Smoking Stands, a-b
6. Pin Cushions and Spool Holder, a
7. Chess Men, a-a'
V. Spiral Turning
1. Pedestal, (Single) a-a', (Double) b
2. Reading Lamps, (Single) a-a'-a'' (Double) b-b'
3. Magazine Holder, a-a'

CHAPTER I

INTRODUCTORY

Wood turning has had a definite place in the commercial world for a great many years. It is used in various forms in making furniture and furniture parts, building trim, tool parts, toys, athletic paraphernalia and many other useful and beautiful articles in common use.

When properly taught in the schools it is one of the most valuable types of instruction. It appeals to pupils more than any other type of manual work, as it embodies both the play and work elements. It is very interesting and fascinating and, in the hands of a skilled instructor, is readily correlated with other work.

Wood turning gives a pupil preliminary experience necessary in pattern making and machine shop work. It brings into play the scientific element by demonstrating the laws governing revolving bodies. In bringing the chisel into contact with the revolving surface, the mathematical principle of the "point of tangency" is illustrated. Excellent tool technique is developed in wood turning as on the exactness of every movement depends the success of the operator, and any slight variation will spoil a piece of work. This brings in a very close correlation of the mental and motor activities and also gives the student an opportunity for observing and thinking while at work. When his tool makes a "run" he must determine the reason and figure out why a certain result is obtained when the chisel is held in a given position. Certain cuts must be fully mastered, and it takes a good deal of experience and absolute confidence in one's self in manipulating the tools before it is possible to attempt skilful work. If scraping is allowed the educational value of the work is lost.

In wood turning a vast field for design and modeling is opened, and art and architecture can be correlated. The pupil will see for himself the need of variety in curves and must use his judgment in determining curves that are so harmonious and pleasing that they will blend together. If properly taught the beauty in the orders of architecture can be brought out in the making of the bead, fillet, scotia, cove, etc.

A feeling of importance is excited in a boy when he sees his hands shaping materials into objects of pleasing form. Wood turning properly taught awakens the aesthetic sense and creates a desire for the beautiful. The boy or man who has learned to make graceful curves and clean-cut fillets and beads will never be satisfied with clumsy effects which are characteristic in cheap commercial work, made only to sell.

Success in turning depends on the following:

1. Care of lathe, tools, selection of materials.
2. Study of the scientific elements of--
a. Revolving bodies.
b. Points of tangency.
c. Study of results by reasoning and observing.
3. Development of technique and exactness.
4. Correlation of mental and motor activities.

CHAPTER II

THE LATHE

The sizes of turning lathes are given as 10", 12", etc. These figures denote the diameter, or size, of the largest piece of work that can be turned on them. The measurement is taken from the center point of the live center to the bed of the lathe (usually 5" or 6") and is one-half the diameter of the entire circle. The length of a lathe is determined by the length of a piece of work that can be turned. This measurement is taken from the points of the live and dead centers when the tail stock is drawn back the full extent of the lathe bed. Fig. 1 shows a turning lathe with sixteen principal parts named. The student should learn the names of these parts and familiarize himself with the particular function of each.

CARE OF THE LATHE

The lathe should be oiled every day before starting. At the end of the period the lathe should be brushed clean of all chips and shavings, after which it should be rubbed off with a piece of waste or cloth to remove all surplus oil. All tools should be wiped clean and put in their proper places. If a student finds that his lathe is not running as it should, he should first call the attention of the instructor to that fact before attempting to adjust it; and then only such adjustments should be made as the instructor directs.

SPEED OF THE LATHE

The speed of the lathe should range from 2400 to 3000 revolutions per minute when the belt is on the smallest step of the cone pulley. At this speed stock up to 3" in diameter can be turned with safety. Stock from 3" to 6" in diameter should be turned on the second or third step, and all stock over 6" on the last step. The speed at which a lathe should run depends entirely upon the nature of the work to be done and the kind of material used. Pieces that cannot be centered accurately and all glued-up work with rough corners should be run slowly until all corners are taken off and the stock runs true.

At high speed the centrificial force on such pieces is very great, causing the lathe to vibrate, and there is a possibility of the piece being thrown from the lathe thus endangering the worker as well as those around him. After the stock is running true the speed may be increased.

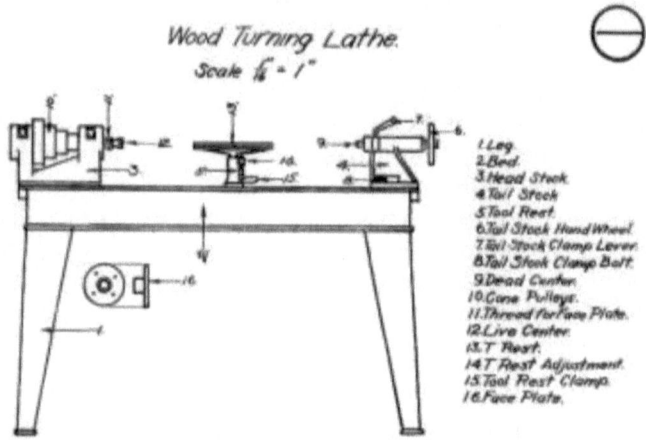

Fig. 1. - Wood Turning Lathe

TO FIGURE THE DIAMETER OF PULLEYS

Suppose a motor runs 1500 R.P.M. and is fitted with a 4" pulley. Suppose also, a main shaft should run 300 R.P.M.

Then, 1500 : 300 :: x : 4;
Or, 300x = 6000,
x = 20, or the diameter of the large pulley on the main shaft.

Suppose again that a line shaft runs 300 R.P.M., and a counter shaft 600 R.P.M. The counter shaft has a pulley 4" in diameter. The pulley on the line shaft must then have a diameter of 8".

300 : 600 :: 4 : x;
Or, 300x = 2400,
x = 8"

Suppose the cone pulley on the counter shaft runs 600 R.P.M.; a lathe spindle runs 2200 R.P.M., when connected with the small cone pulley which has a diameter of 3". The large cone pulley has then a diameter of 11".

600 : 2200 :: 3 : x
Or, 600x = 6600;
x = 11"

RULES FOR FINDING THE SPEEDS AND SIZES OF PULLEYS

1. To find the diameter of the driving pulley:

Multiply the diameter of the driven by the number of revolutions it should make and divide the product by the number of revolutions of the driver. (20 x 300 = 6000; 6000 ÷ 1500 = 4"--diameter of motor pulley.)

2. To find the diameter of the driven pulley:

Multiply the diameter of the driver by its number of revolutions and divide the product by the number of revolutions of the driven. (4 x 1500 = 6000; 6000 ÷ 300 = 20"--diameter of the driven pulley.)

3. To find the number of revolutions of the driven pulley:

Multiply the diameter of the driver by its number of revolutions and divide by the diameter of the driven. (4 x 1500 = 6000; 6000 ÷ 20 = 300--revolutions of driven pulley.)

POINTS ON SETTING UP LATHE AND SHAFTING

The counter shaft should be about 7' above the lathe. A distance of 6' from the center of the shaft to the center of the spindle is sufficient. In setting a lathe or hanging a counter shaft it is necessary that both be level. The counter shaft must be parallel to the line shaft. When the counter shaft is in position a plumb bob should be hung from the counter shaft cone to the spindle cone; the lathe should be adjusted so that the belt will track between the two cone pulleys. The axis of the lathe must be parallel to that of the counter shaft. The lathe, however, need not be directly beneath the counter shaft as the belt will run on an angle as well as perpendicular.

CHAPTER III

WOOD TURNING TOOLS

A wood turning kit should consist of one each of the following tools. Fig. 2 shows the general shape of these tools.

1¼" Gouge
¾" Gouge
½" Gouge
¼" Gouge
1¼" Skew
¾" Skew
½" Skew
¼" Skew
⅛" Parting Tool
½" Round Nose
¼" Round Nose
½" Square Nose
¼" Square Nose
½" Spear Point
½" Right Skew
½" Left Skew
Slip Stone with round edges
6" Outside Calipers
6" Inside Calipers
8" Dividers
12" Rule
½ pt. Oil Can
Bench Brush

GRINDING AND WHETTING TURNING TOOLS

Skew Chisel

The skew chisel is sharpened equally on both sides On this tool the cutting edge should form an angle of about 20° with one of the edges. The skew is used in cutting both to the right and to the left, and therefore, must be beveled on both sides. The length of the bevel should equal about twice the thickness of the chisel at the point where it is sharpened. In grinding the bevel, the chisel must be held so that the cutting edge will be parallel to the axis of the emery wheel. The wheel should be about 6" in diameter as this will leave the bevel slightly hollow ground. Cool the chisel in water occasionally when using a dry emery. Otherwise the wheel will burn the chisel, taking out the temper; the metal will be soft and the edge will not stand up. Care should be exercised that the same bevel is kept so that it will be uniformly hollow ground. The rough edge left by the emery wheel should be whetted off with a slip stone by holding the chisel on the flat side of the stone so that the toe and heel of the bevel are equally in contact with it. Rub first on one side and then on the other. The wire edge is thus worn off quickly as there is no metal to be worn away in the middle of the bevels. The chisel is sharp when the edge, which may be tested by drawing it over the thumb nail, is smooth and will take hold evenly along its entire length. If any wire edge remains it should be whetted again.

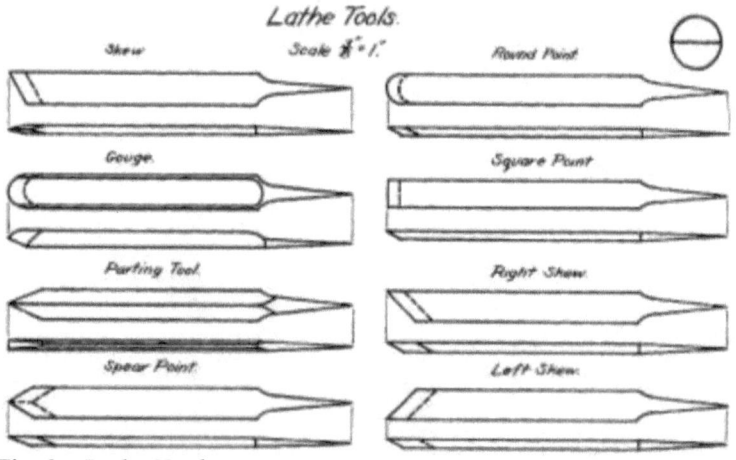

Fig. 2. - Lathe Tools

Gouge

The gouge used in wood turning is beveled on the outside and is ground so that the nose is approximately semi-circular in shape. The tool is a combination of the round nose chisel and the ordinary gouge. The bevel should extend well around to the ends so that the cutting edge extends to each side. This is necessary to avoid the abrupt corners which would be present if the nose were left straight across as in the ordinary wood-working gouge. In making shearing cuts the round nose permits the tool to be rolled to the side to avoid scraping the work. The length of the bevel should be about twice the thickness of the blade at the point where the sharpening begins.

The sharpening of a gouge for turning is rather difficult for the average student. The ordinary gouge which has a square nose may be beveled by merely turning it half way around and back again. In working out the round nose of a gouge for wood turning, it is necessary that the handle be swung from one side to the other while, at the same time, the chisel is revolved to cut the bevel evenly. It is sometimes necessary to allow some pupils to use the side of the emery wheel in sharpening the gouge. This kind of grinding, however, does not leave the tool hollow ground as when the face of the wheel is used.

To complete the sharpening the rough edge is worked smooth on a slip stone, the cross section of which is wedge-shaped and the edges of which are rounded. The toe and heel of the beveled side of the gouge are brought into contact with the flat side of the stone. As the sharpening proceeds the wire edge is worked to the inside of the gouge. The rounded edge of the stone is then placed inside the gouge and is worked back and forth until the rough edge disappears. Great care must be taken not to bevel the inside of the gouge when whetting with the round edges of the stone, as the result will be the same as with an ordinary chisel or plane bit.

Parting Tool

The parting tool is sharpened on both sides. This tool differs from the ordinary chisel in that it is between ⅝" and ¾" thick and only about ⅛" wide at the widest point, which is in the center of its entire length. The bevels must meet exactly at the center, or the widest point, and should make an angle of about 50° with each other. If the bevels do not meet at the widest point the tool will not clear, and the sides will rub against the revolving stock; the tool will be burned and will thus lose its temper. The bevel should be hollow ground slightly as then comparatively little metal need be removed when whetting.

Scraping Tools

The round nose, square nose, spear point, right skew and left skew are scraping tools, used chiefly in pattern work and sometimes in face-plate work. They are sharpened on one side only, and the bevel is about twice the thickness of the chisel at the point where sharpened. These tools should be slightly hollow ground to facilitate the whetting. Scraping tools become dull quite easily as their edges are in contact with the wood almost at right angles. After sharpening, the edges of these tools may be turned with a burnisher or the broad side of a skew chisel in the same manner that the edge of a cabinet scraper is turned though not nearly to so great a degree. This will help to keep the tool sharp for, as the edge wears off, the tool sharpens itself to a certain extent. The chisel is of harder material than a cabinet scraper so that it will not stand a great amount of turning over on the edge. Small pieces will be broken out, unless a flat surface is rubbed against the edge at a more acute angle than was used in the whetting. If a narrow burnisher is used, pieces are more likely to be broken out from the sharp edge and thus make the tool useless.

CHAPTER IV

SPINDLE TURNING

Spindle turning is the term applied to all work done on a lathe in which the stock to be worked upon is held firmly between the live and dead centers. There are two methods in common use in wood turning: first, the scraping or pattern-makers' method; and second, the cutting method. Each has its advantages and disadvantages, but it is necessary that both be learned in order to develop a well rounded turner. Care should be exercised, however, that each method be used in its proper place. The first is slower, harder on the cutting edge of tools, and less skill is required to obtain accurate work; the second is faster, easier on the cutting edge of tools, and the accuracy of results obtained depends upon the skill acquired. As skill is the one thing most sought for in high school work, the use of the cutting method is advocated entirely for all spindle turning and, with but few exceptions, for face-plate and chuck turning.

TO CENTER STOCK

If the wood to be turned is square or rectangular in shape the best way to locate the center is to draw diagonals across the end of the stock. The point of intersection locates the center.

CLAMPING STOCK IN THE LATHE

Take the live center from the spindle and with a wooden mallet drive the spur deep into the wood. Never drive the wood onto the live center while in the spindle because serious injury may be done the machine by such practice. When extremely hard wood is being used, it is a good practice to make saw cuts along the diagonal lines and bore a hole at the intersection, thus allowing the spur to enter the wood more freely. Oil the other end of the wood while holding it in a vertical position, and give the oil a chance to penetrate into the wood. Then replace the live center by taking the stock and center and forcing it into the spindle by a sudden push of the hand. The tail stock is then moved about ½" to 1" from the end of the piece to be turned, having the tail spindle well back in the tail stock. The tail stock is then clamped to the lathe bed. Turn the tail stock hand wheel until the wood is held firmly. Work the cone pulley by hand at the same time, so that the cup or dead center will be forced deeply into the wood, so deeply that the live center will not continue to turn. Now turn the dead spindle back until the live spindle begins to turn freely and clamp the dead spindle fast.

Fig. 3

ADJUSTING THE TOOL REST

Horizontally the tool rest should be set about ⅛" from the farthest projecting corner of the wood and should be readjusted occasionally as the stock diminishes in size. The vertical height varies slightly according to the height of the operator. It is even with the center of the spindle for a short person; 1/8" above for a medium person; and 1/4" above for a tall person. So long as the stock is in its square form the tool rest should never be adjusted while the machine is in motion as there is danger of the rest catching the corners and throwing the stock from the machine. Also see that everything is clamped tight before starting the lathe.

POSITION OF THE OPERATOR

The operator stands firmly on the floor back far enough from the lathe to allow him to pass the tools from right to left in front of his body without changing the position of the feet. It may be found convenient to turn slightly, bringing the left side of the body a little closer to the lathe. In no case, however, should the tools be brought in contact with the body as the cutting operation from right to left should be accomplished by a movement of the arms alone and not the swaying of the body. (Fig. 3.)

HOLDING THE TOOLS

All tools should be held firmly but not rigidly. The right hand should grasp the handle at the extreme end for two reasons: first, to give as much leverage as possible so that the tool will not be thrown from the hands in case it should catch in the wood; second, a slight wavering of the hand will not cause as much variance in the cuts as when held closer up to the rest. The left hand should act as a guide and should be held over the tool near the cutting edge. The little finger and the back part of the palm of the hand should touch the tool rest thus assuring a steady movement. The left hand should not grasp the tool at any time. (Fig. 3.)

USE OF THE TOOLS IN SPINDLE TURNING

The correct use of the various tools used in spindle turning will be explained in detail as the steps are worked out in the sequence of operations on the exercises in Section A-I.

CHAPTER V

TOOL PROCESSES IN SPINDLE TURNING

Exercise A-I--1-a. Straight Cuts

1. THE ROUGHING CUT (LARGE GOUGE).

FIG. 4. Place the gouge on the rest so that the level is above the wood and the cutting edge is tangent to the circle or surface of the cylinder. The handle should be held well down.

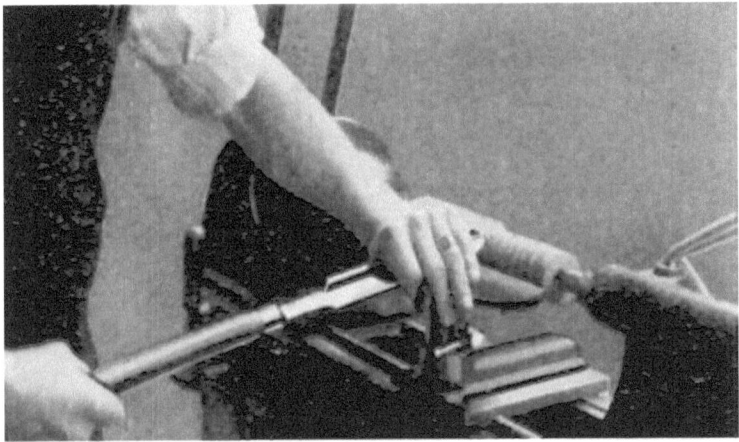

Fig. 4.

Roll the gouge over slightly to the right so that it will make a shearing cut instead of a scraping cut. This rolling of the tool will also throw the chips from the operator.

Then lift the handle slowly, forcing the cutting edge deep enough into the wood to remove all or nearly all of the corners, at the end of the work which is being turned. This cut is begun about ¾" from the dead center end. Work back another ¾", moving toward the live center and make a second cut, and so on until the entire length of the cylinder is gone over. This method of removing corners should always be followed to avoid any possibility of breaking a large sliver from the stock, with consequent danger to the worker.

The tool may then be worked from one end to the other, getting a fairly-smooth, regular surface, slightly above the diameter required. However, do not begin on the very edge of the cylinder end. It is better to begin about 2" from one end and work to the other, and then reverse and work back.

The tool should also be held at a slight angle to the axis of the cylinder, with the cutting point always in advance of the handle.

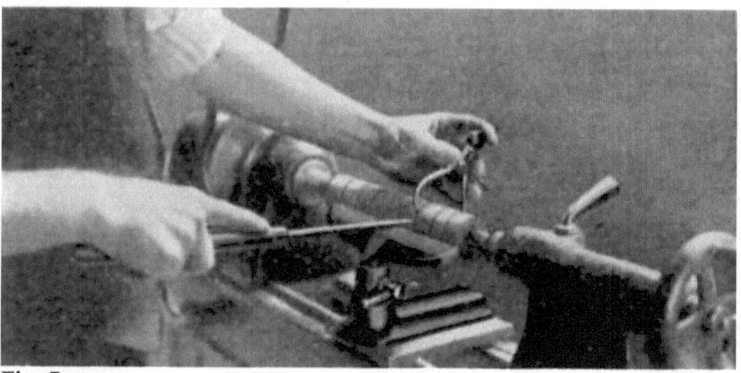

Fig. 5.

2. THE SIZING CUT (SMALL GOUGE). FIG. 5.

Set the calipers to the required diameter of the cylinder.

With a small gouge held in the right hand scrape grooves about 1" apart, holding the calipers in the left hand perpendicular to the cylinder and measuring the cuts as they are made. The scraping should continue until the calipers will pass easily over the cylinder. It will be well while scraping to work the handle of the gouge a little from side to side so that the nose has more clearance. This will prevent the piece which is being turned from chattering or vibrating.

The calipers will be slightly sprung by coming in contact with the revolving stock but this error in diameter will be removed by the finishing cut which removes these marks from the finished cylinder.

3. THE SMOOTHING CUT (LARGE SKEW).

FIG. 6. Lay the skew chisel on the rest with the cutting edge above the cylinder and at an angle of about 60° to the surface.

Slowly draw the chisel back and at the same time raise the handle until the chisel begins to cut about ¼" to ⅜" from the heel. The first cut is begun from 1" to 2" from either end and is pushed toward the near end. Then begin at the first starting point and cut toward the other end. One should never start at the end to make a cut as there is danger that the chisel will catch and cause the wood to split or that the chisel will be torn from the hands.

The first cut takes off the bumps and rings left by the gouge, and takes the stock down so one can just see where the scraping to size was done. Then take the last cut and remove all traces of these, leaving the cylinder perfectly smooth and of the required diameter at each end. Test the cylinder for accuracy with a straight edge.

Fig. 6.

4. TESTING FOR SMOOTHNESS. In testing for smoothness place the palm of the hand, with the fingers extended straight, lightly on the back of the cylinder opposite the tool rest. This position will avoid any possibility of the hand being drawn in between the cylinder and the rest.

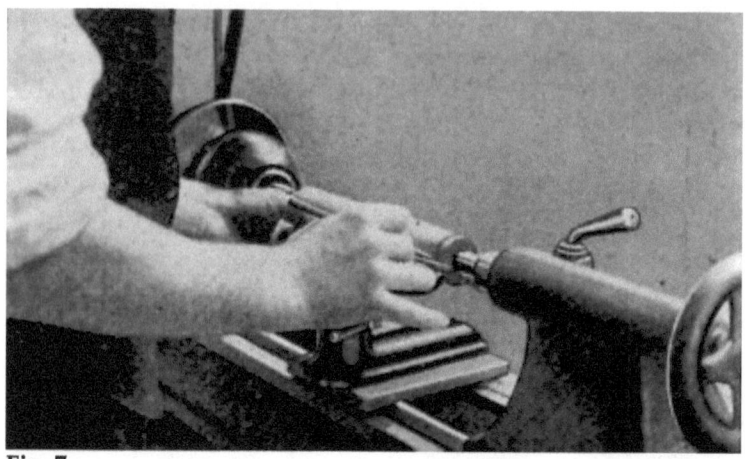

Fig. 7.

5. MEASURING FOR LENGTH (RULE AND PENCIL). FIG. 7. Hold the back edge of the rule in the left hand and place it on the tool rest so that the front edge is almost in contact with the revolving cylinder.

With a sharp pencil mark off the required length, starting from the dead center end. The first mark should be just far enough in on the cylinder to insure cutting past the point of the dead center. This will leave all surplus stock at the live center end where it is needed, because, if not enough stock is left at this end, there is danger of striking the live center spur with the tool and of injuring the chisel and perhaps the work.

In case several measurements are to be made, as in some of the following exercises, the rule should not be moved until all are marked. This will insure more accurate work than if the rule be changed several times.

6. SQUARING ENDS (SMALL SKEW AND PARTING TOOL). FIG. 8. This operation is done with the toe or acute angle of the ½" or ¼" skew chisel.

Place the chisel square on the tool rest. Swing the handle out from the cylinder so that the grind, which forms the cutting edge, next to the stock is perpendicular to the axis of the cylinder. The heel of the chisel is then tipped slightly from the cylinder in order to give clearness. Raise the handle and push the toe of the chisel into the stock about ⅛" outside the line indicating the end of the cylinder. Swing the handle still farther from the cylinder and cut a half V. This will give clearance for the chisel point and will prevent burning. Continue this operation on both ends until the cylinder is cut to about 3/16" in diameter.

The remaining ⅛" is then removed by taking very thin cuts (about 1/32") holding the chisel as first stated. After each cut is made the end should be tested for squareness by holding the edge of the chisel over the end of the cylinder.

Fig. 8.

This is an easy cut after it is mastered, but is one of the hardest to learn. Should the operator lose control of the tool and allow any part other than the point to touch the cylinder, a run or gashing of the wood will be caused.

In large cylinders where considerable stock has to be cut away in order to square the ends, time will be saved by sizing the ends down with the parting tool to within ⅛" of the desired line, leaving enough stock at the base of the cuts to still hold the cylinder rigid while cutting on the ends.

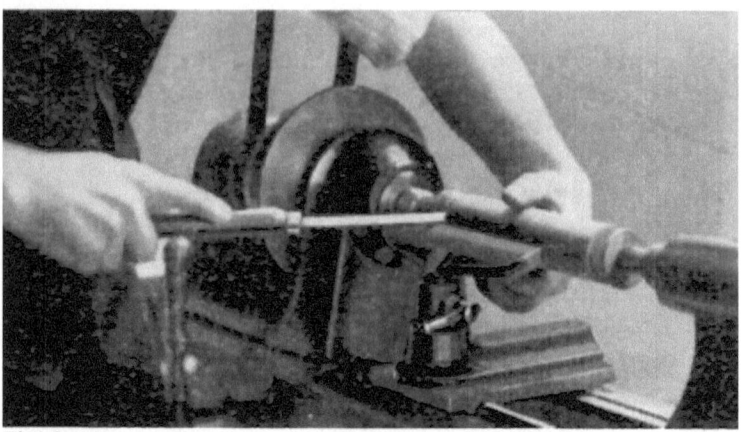

Fig. 9.

For this operation hold the parting tool on the rest with the cutting edge parallel to the axis of the cylinder and the lower grind tangent to the cylinder. Lift the handle and force the cutting edge into the wood; at the same time push the chisel forward to keep it at the proper tangency.

7. CUTTING OFF (SMALL SKEW). FIG. 9.

After both ends have been squared cut away stock, at both ends, to leave just enough to hold the cylinder from separating from the waste ends.

With the chisel held in the right hand in the same position as in squaring the ends, and the fingers of the left hand around the stock to catch it, slowly force the point of the chisel into the stock at the live center end, until it is cut free and the cylinder stops in the operator's hand. Too much pressure should not be used in this operation or it will cause the cylinder to twist off instead of being cut, and will leave a ragged hole in the end.

The dead center end, which has been scored heavily before cutting off at the live center, is then removed by holding the grind of the chisel flat on the end of the cylinder. The latter is revolved by hand until the stock is cut away.

Exercise A-I--2-a. Shoulder Cuts

1. Turn a cylinder to the largest diameter required.

2. Lay off measurements with rule and pencil.

3. With the gouge (where space permits) or the parting tool (in narrow spaces) rough out surplus stock, keeping 1/16" away from the lines indicating shoulders.

4. Caliper to the diameter of the second step.

5. The shoulders are cut down as described in "Squaring Ends, Step 6, Straight Cuts."

6. The new diameter or step is then trued up with a skew chisel in the same manner as a cylinder; except that in nearing the shoulder the chisel is pushed up on the cylinder until the heel, which is the only part that can be worked into the corner, becomes the cutting point. Fig. 10. In very narrow steps it will be advisable to use the heel entirely as a cutting point.

In spaces between shoulders, too narrow to permit the use of the skew chisel, very effective work can be accomplished by slightly tipping the parting tool sideways to allow a shearing cut to be taken with the cutting edge.

7. Where several steps are required on the same cylinder, each successive one is worked out as above described.

Note:--All preliminary steps in working stock to size, laying of dimensions, etc., in preparation for the exercise in hand, will be omitted in the following exercises:

Exercise A-I--3-a. Taper Cuts

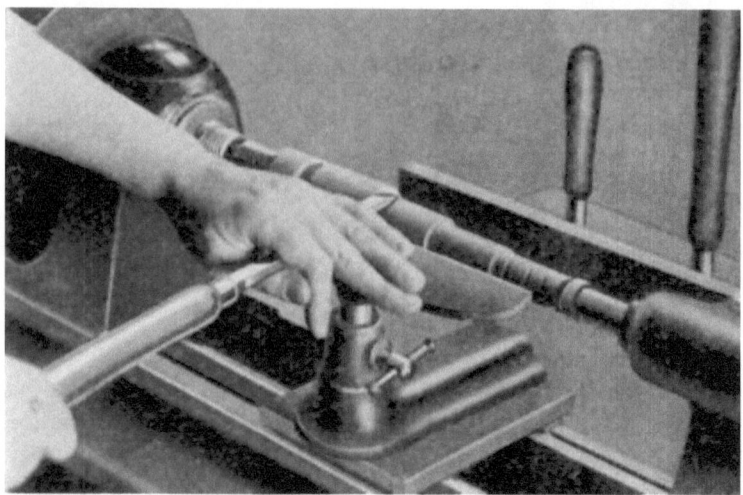

Fig. 10.

Calipering for New Diameters. For all diameters on tapers the calipers should be set 1/16" larger than the desired measurement in order to avoid working under size in the finishing cut which removes all caliper marks.

If the taper runs to the extreme end of the cylinder, as in Plate A-I--3-a, a parting tool should be used, instead of a gouge, to take off a very thin shoulder.

If the taper forms an internal angle as in Plate A-I--3-b, a gouge is used as in Step 2--Sizing Cut--Plate A-I--1-a.

In other cases where tapers connect with straight cylindrical shoulders it is best to turn the shoulders to size before working the tapers.

In cutting a long straight taper the skew chisel is used, much in the same manner as in ordinary cylinder work, except that at the start of each cut the heel must be the cutting point. This will avoid any chance of the chisel catching and drawing back and thus gouging the wood beyond the starting point. As soon as the cut is well under way the chisel may be pushed up on the cylinder so that the cutting point is a little above the heel. All cuts should be made from the highest point on the cylinder to the lowest and thus cut across the grain of the wood.

In making the cut, care should be taken to see that the chisel is not tipped to a greater angle than that of the taper wanted. Should that be done a hollow, or dished out, taper is sure to be the result instead of a straight one.

Exercise A-I--4-a. V Cutting

In cutting V's a small skew is almost always used and the cutting is done with the heel.

Place the chisel square on the tool rest so that the cutting edge is perpendicular to the axis of the cylinder. Draw the chisel back and raise the handle so that the heel is driven into the wood, thus scoring it. This cut should not be too deep or the chisel will burn. This scoring should be at the exact center of the V cut.

Swing the handle a little to the right and at the same time tip the chisel so that the grind, which forms the cutting edge, is at an angle of about 45° with the axis of the cylinder. The handle is then raised at an angle of 45° bringing the heel down to make a good cut. The chisel is then swung to the other side and a similar cut is taken. These cuts are continued, together with the center scoring, until quite close to the pencil marks. Test the angle before the finishing cut is taken.

It will be found best to have the V slightly greater than 90° at the base until the final cut is made, at which time it can be trued up.

The V should be tested with the square end of a rule. The cylinder should not be in motion while testing.

When angles other than 45° are cut, the cutting edge of the chisel should be tipped so that it is parallel or nearly so to the side of the cut desired.

A-I--5-a. Concave Cuts

The concave cuts as a rule will give the pupil considerable trouble at first owing to the fact that the grind, which forms the cutting edge and which must be held perpendicular to the cylinder at the start, is on the under side of the tool and cannot be seen. However, as soon as the correct angle of the tool is located, the cut will be found as easy as any. Concaves are usually made with a medium sized gouge either the ½" or ¾".

Place the gouge on the rest with the grind or cutting edge well above the wood. The tool is then rolled on its side so that the grind at the cutting point, which is on the lip of the gouge well below the center, is perpendicular to the axis of the cylinder. Fig. 11.

Slowly raise the handle to force the gouge into the wood. As soon as the gouge has taken hold, the tool is forced forward and upward by a slight lowering of the handle, while at the same time it is rolled back toward its first position. Care should be taken not to roll the chisel too fast or a perfect arc will not be cut.

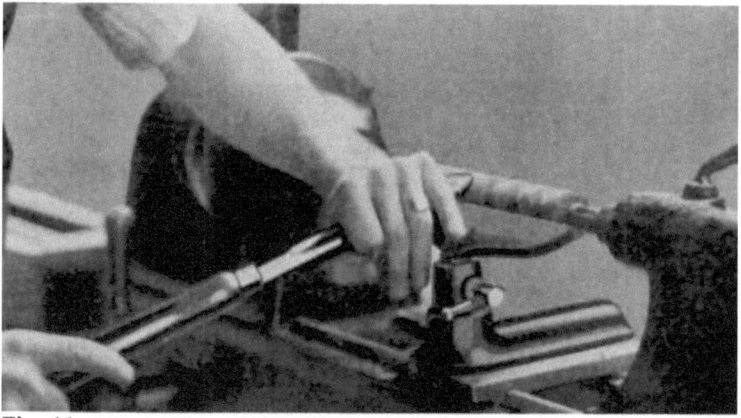

Fig. 11.

By this triple action the grind, which comes in contact with the surface of the curve, forces the lip sidewise and cuts one quarter of a circle. Reverse the position of the gouge and cut from the other side in the same manner to form the other half of the semi-circle. The cutting should always stop at the base of the cut as there is danger that the tool will catch when cutting against the grain of the wood on the other side. Repeat this operation until within about 1/16" of the required size. At the end of each successive cut the tool should have been forced far enough forward and upward to bring the grind or nose of the chisel well out on top of the cut. Fig. 12.

The exact depth of the concave is then calipered in the usual manner as described before. A finishing cut is then taken after the cut has been tested with a templet.

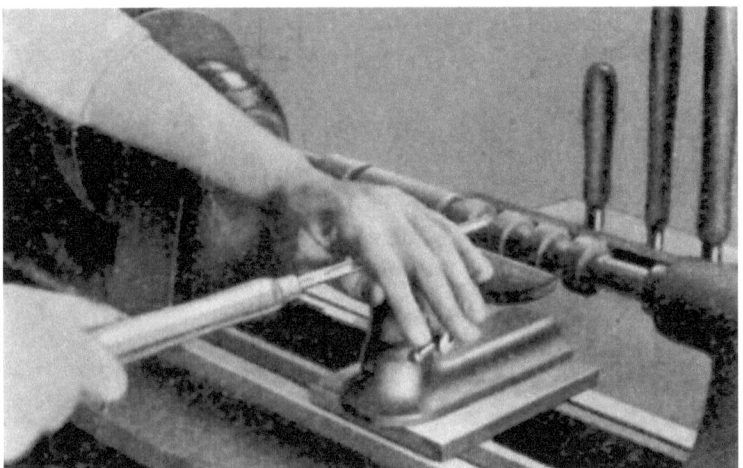

Fig. 12.

A-I--6-a. Convex Cuts

The convex cut, or Bead as it is usually called, is generally considered the hardest cut to make.--The cut is made with the heel of a small skew chisel, usually the ¼" or ⅛".

After the cylinder has been marked off, rough out all stock between the beads with a parting tool. The base of the cuts is finished the same as described in Plate A-I--1-a, for shoulder cutting. With a sharp pencil mark the center of each bead to be made. This line is the starting point for all cutting.

Place the chisel on the rest, with the cutting edge above the cylinder and the lower grind tangent to it. Draw the chisel back and raise the handle to bring the heel of the chisel in contact with the cylinder at the line indicating the center of the bead. The chisel is then moved to the right (if cutting the right side of the bead); at the same time the chisel is continually tipped to keep the lower grind tangent to the revolving cylinder and also to the bead at the point of contact. Fig. 13. This cut is continued until the bottom of the bead is reached.

It is well in turning a series of beads to work the same side of all before reversing to the other side.

Note:--The same principles employed in this exercise are also used in working out long convex curves such as are found in chisel handles, mallet handles, etc. The only exception is that in most cases the point of contact need not be the heel of the chisel but higher up as in ordinary straight work.

A-I--7-a--Combination Cuts

These exercises are so designed as to include one or more of each of the foregoing cuts. The student here is given an opportunity of combining these cuts into one finished product.

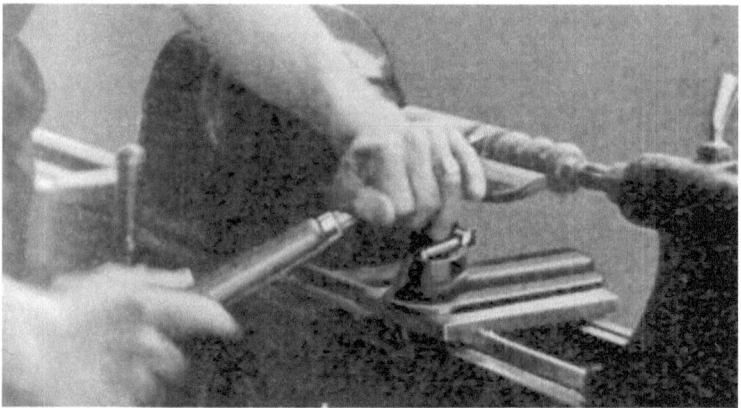

Fig. 13.

An analysis of the exercise chosen should be made to determine which of the various cuts should be made first, second, etc., in order to produce the exercise in the shortest time and with the least amount of tool manipulation.

After the student has mastered these cuts with a certain degree of skill and accuracy, he is ready to apply them in working out various models in Section II.

A-II--1-a. Chisel Handles

At this point it is well to state that the small end of all work should be turned at the dead center. In the case of chisel handles the socket or ferrule end is at the dead center where the stock can safely be cut away to permit the fitting of the ferrule or the socket.

After the stock is turned to a cylinder of the largest dimension desired, the taper, for the socket chisel, should be turned first and fitted to the chisel in which it is to be used. Then the rest of the handle is worked out. Ferrules should also be fitted in the same manner. A drive fit should be used for all ferrules.

A-II--2 and 3. Mallets and Gavels

The biggest source of trouble in these models is getting the handles to fit true. This is caused by not getting the hole in the head straight.

Turn the head to a cylinder 3/16" larger than the finished dimension. Then bore the hole perpendicular to the axis as near as possible, either by leaving it between the lathe centers or by placing it in a vise. The handle is then fitted into the head. A snug fit is necessary. If one side "hangs" or is lower than the other the centers are moved sufficiently to correct it. The head is then turned to exact size and finished.

A-II--7. Vise Handles.

Turn the spindle with the solid head to dimensions. Bore a hole through a 1¼" square block and fit the block snugly to the end of the spindle. Turn this block to the same dimensions as the other head. This method will save chucking the second head and is much quicker.

CHAPTER VI

OVAL TURNING

Oval work as a problem in turning will be found to be a very good one as well as interesting to the pupil. It brings in the principle of the oval as used in ordinary shop practice; (arcs from points on the major and minor axes). For thick heavy ovals the off-centering is very slight, while for long, thin ones the off-centering is greater. The measurements given on Plates A-III--1-a, b and A-III--2-a, b will give a good idea of approximate distances to be used.

While the tool operations are much the same as in other spindle turning there is one notable difference. The design must be worked out by eye, because of the nature of the work no caliper measurements can be made for depth of cuts.

To get the best results the stock of oval turning should be cut square or slightly rectangular in cross-section and about 3" longer than the model to be made. The thickness of the stock should be about ⅛" greater than the major axis of the oval wanted.

The centers are located in the usual manner after which perpendicular lines are drawn from the sides, passing through the points of the centers. From the ends of one of these, perpendicular lines are extended lengthwise of the stock (on opposite sides) meeting the corresponding perpendicular at the other end of the stock. These lines form the ridge of the oval. On the other perpendiculars, the points for off-centering are laid off, measuring the required distance on both sides of the center point.

With a ⅛" drill bore holes ¼" deep at each of the off-centering points as well as the original center. This will insure the lathe centers penetrating the stock at the proper point. The stock is then placed in the lathe, using two corresponding off-center points as centers.

With the lathe running at third speed turn down the stock to the horizontal line forming the ridge of the oval, excepting for a distance of about 1¼" at the ends. The stock at the ends is necessary for the off-centering and, if cut away, will spoil the centering for the other side, especially at the live center end. The stock is then changed to the other off-center points and the second side is cut down to the line.

All measurements are then laid out and the design is cut, changing the stock in the lathe when necessary. Care should be taken that the sharp ridge left on the work forms a straight line the full length of the stock. After the design has been finished, the stock is centered on the true center and a very thin cut is taken the full length of the object to remove the sharp corners. The model is then sandpapered while the lathe is running very slowly.

CHAPTER VII

DUPLICATE TURNING

Under the head of duplicate turning have been classified only such models as clearly indicate the necessity of making two or more articles to complete the model or set of articles desired. But it is not intended to convey the idea that other models may not be made in duplicate as in many cases it is very desirable and even preferable that they should be made that way. Whatever the problem may be the suggestions offered at this point may be applied effectively.

Whenever two or more models are to be made identically alike there are always two possibilities of inaccuracies that will render the work dissimilar: First, inaccurate measuring both for length and points of new diameters and also on the new diameters themselves; second, a variation in the curved surfaces either on long convex or concave cuts.

The first difficulty can be overcome to a great extent by the use of a measuring stick. This stick should be made of any soft wood. It should be straight on one edge and about the thickness of an ordinary rule. On the straight edge lay off very carefully measurements for length, shoulders, beads, concaves and all points where calipering for new diameters will be necessary. Insert at each point measured a small brad which has been sharpened at both ends, leaving the end protrude about ⅛". Care should be taken that all brads protrude the same distance.

After the stock has been turned to the largest diameter, the stick is held in the position of the rule while measuring and the points are forced against the revolving cylinder, thus scoring it. This stick can be used as many times as the model is to be made and the measurements will always be the same.

To avoid dissimilar curves it is well to cut out a full sized templet of the model to be made. This templet can be made of any thin, stiff material, preferably light sheet iron. In some cases it will be necessary to make the templet in several pieces in order to help facilitate the tool operations.

The use of this templet will not only be a help to getting all curved surfaces the same, but will also check up on the various new diameters on the model. The cylinder should never be in motion while the templet is being used.

CHAPTER VIII

FINISHING AND POLISHING

To get a high and lasting polish on wood, the work must be first sanded so as to be perfectly smooth. In addition to this, open grained wood, such as oak, must be properly filled with a wood filler. If properly sharpened tools have been used very little sanding is required, and then worn sandpaper should be used as it does not cut into the work as new paper cuts. Remember sandpaper is not to be used as a tool in cutting down stock when working to dimensions. In using old sandpaper run the lathe at a moderate speed to avoid burning the wood, especially on square or round fillets. Keep the edges of the work sharp and do not wear them round. In using new sandpaper use a fine grit (00 or 0) and move the paper from one end of the work to the other slowly, so that no scratches result on the surface of the work.

The work may be finished by one of two methods. In the first method as in finishing ordinary cabinet work, the pieces should be stained and filled. In applying filler, run the lathe at the slowest speed after the material has dried sufficiently to rub into the pores of the wood. If the highlights are to be brought out, as in the case of oak, stain and then give a light coat of shellac, and apply the filler after the shellac is dry. The shellac keeps the dark filler from staining the flakes of the oak darker, and the pores of the wood fill in as before. The pores become darker than the flakes, and at the same time a smooth surface is produced. After the filler has hardened the wood may be waxed or varnished.

The second method, or French polishing, is rather difficult to apply and requires a little skill. A close grained wood, like maple, will be found more satisfactory for the beginner. An open grained wood may be filled in the ordinary way, or the grain may be filled by rubbing into the pores of the wood a combination of shellac, rotten stone or pumice, oil and alcohol. Rotten stone is used for dark wood and pumice is used for light wood. The wood may be left in the natural or stained as in the first method. The mixture of shellac, rotten stone, oil and alcohol, is applied to the work with a pad made of cotton waste, wrapped in cheese cloth to keep it from sticking to the work. It should be about 1½" in diameter and ½" thick. Hold the pad over the mouth of a bottle of shellac and tip the bottle so that the shellac comes in contact with the pad. The shellac will remain clean in a bottle and will be handy. The mouth of the shellac bottle should be about 1" in diameter and should be dipped once. Do likewise with a bottle, having a mouth ½" in diameter, containing alcohol. This should be dipped twice allowing the alcohol to dilute the shellac. Then drop on a couple of drops of oil and rub over the pad evenly; this aids in distributing the shellac properly and keeps the pad from sticking to the work. A bottle may also be used for this. For the rotten stone use a pepper shaker so that it may be sifted on the work as needed.

When the mixture has been applied to the pad, hold the pad against the work lightly at first, until most of the moisture has been worked out of it, and then gradually increase the pressure until the pad is almost dry. In putting on the first coat, use more shellac and alcohol and just enough oil at all times to prevent the pad from sticking to the work. However, the pad should not contain as much shellac that it can be squeezed out with the fingers. When the pad is dry, another mixture is applied, and where open grained wood is used, rotten stone, or pumice stone, is sprinkled on the work to gradually fill up the pores and to build up a smooth surface. Run the lathe at a low speed, depending on the size of the piece that is being polished. Allow the first coat to dry before applying a second coat for, if too much is put on at any one time, the heat generated in the rubbing will cause the shellac to pull, and it will form rings by piling up. These rings may be worked out in two ways, either by a slight pressure of the pad on the rings or by cutting them with alcohol applied to the pad. If too much alcohol is used it will cut through the shellac and remove what has already been rubbed on. If at any time too much shellac is used it will pile up and form rings. Too much rotten stone will cut down the polish and by absorbing the mixture will leave the pad dry. If too much oil is used the polish will become dull after a day or two.

After the first coat has hardened apply the second, but use less shellac and more alcohol and just enough oil to prevent the pad from sticking. This may be done by dipping the tip of a finger in the oil and spreading it over the pad. The entire mixture should be so that only a dampness can be felt on the pad. As the process goes on less oil and shellac are used. All oil must be removed when applying the last coat, or the piece will lose its polish. All the pores should be filled, and no rings should be on the finished work. Where a natural finish is desired, apply a coat of boiled linseed oil twelve hours before the work is to be polished. This will bring out the grain and will also aid in applying the first coat; no oil need then be used in the first coat.

A great amount of practice and patience is required to get a first class polish. Polishing can only be learned by experience. Correct your troubles in properly proportioning the mixture. Never use too much shellac as it will build up too fast and will not harden, thus causing rings; or it will pull and catch to the pad, thus forming bunches. The purpose of alcohol is mainly to dilute the shellac and to prevent against putting it on the work too fast, but care must be taken not to use too much alcohol to cut the shellac entirely. The oil helps to distribute the shellac evenly, but it must be removed when finishing the last coat, or the polish will not remain. It also helps to keep the pad from sticking to the work.

It is impossible to obtain a polish that will be as lasting and rich by any method other than the one described. For success it is essential to learn the proportions of the mixture and to acquire skill in applying the materials by using exactly the right pressure and the right movement of the pad.

CHAPTER IX

FACE-PLATE AND CHUCK TURNING

Face-plate and chuck turning open an entirely new field of work from that taken up in previous chapters of this book. If handled correctly, it has much greater educational and practical value than cylinder turning. From the practical standpoint the field of work is broader and the models to be made are of much greater value. Aside from this, trade methods and practices can be applied and a broad insight into commercial work can be given the student.

In some details of chuck turning the tool operations already learned can be employed, but for the most part they are entirely different. In order to preserve the educational value of the work as brought out by skill and dexterity in handling tools, it will be necessary to use the cutting method wherever possible. In some instances that method will be impossible, and the scraping method must be used.

METHODS OF FASTENING STOCK

All the work thus far has been on models where the stock worked upon is held between the live and dead centers. In face-plate and chuck turning the work is done at the head stock only and the piece is supported by means of a face-plate, or chuck, that is fastened to a face-plate, which is screwed onto the end of the live spindle. There are three methods of fastening stock to the face-plate, and it depends upon the nature of the exercise or model to be made which method is used.

1. SMALL SINGLE SCREW FACE-PLATE. For all work that does not require deep cutting in the center, such as in towel rings, picture frames, etc., the small face-plate with a single screw should be used.

Note:--Should it be found difficult to keep the block from working loose and turning, it is a good plan to fold a piece of sandpaper, grit side out, and place it between the face-plate and the stock.

2. LARGE SURFACE SCREW FACE-PLATE. For all work that does not require deep cutting on the outside, such as exercises, jewel boxes, etc., as well as all large stock, and all stock from which chucks are to be made, the large face-plate with the surface screws should be used.

3. GLUING TO WASTE STOCK. A block of scrap wood is fastened to a face-plate the same as for a chuck and surfaced off square. The block from which the model is to be made is planed square on one side and glued to the block on the face-plate with a sheet of paper between the two. To separate the model from the chuck, after it is completed, place a chisel on the waste stock, 1/16" back of the glue joint at such a point as will bring the chisel parallel to the grain of the model, and strike lightly with a mallet. This will cause the paper to separate and the model to become free.

This method will be found very convenient epecially on models where the base is to be left straight. It will also be found to save much stock when working with expensive woods.

LATHE ADJUSTMENTS

To get the best results in face-plate or chuck turning there should be no end play in the spindle of the lathe. The spindle should always be tested out, and if any play is found, should be adjusted before attempting any work. It is almost impossible to make a true cut when such a condition obtains.

POSITION OF TOOL REST

For all face-plate and chuck turning the tool rest should be kept as close to the stock as possible, the same as in spindle turning, regardless of the angle it may be set. Vertically, the rest in most cases should be sufficiently below the center of the stock to bring the center or cutting point of the tools used, when held parallel to the bed of the lathe, even with the center of the stock. This last condition will necessitate adjusting the height occasionally when changing from large to small tools.

CHAPTER X

TOOL PROCESSES IN FACE-PLATE AND CHUCK TURNING

B-I--1-a. Straight Cuts

1. ROUGHING OFF CORNERS. (¾" GOUGE.) FIG. 14. The tool rest is set crosswise to the bed of the lathe and parallel to the face of the stock.

Place the gouge on the rest with the handle well down. Roll the gouge to the left until the grind which forms the cutting edge is perpendicular to the stock. The point of contact should be slightly below the center or nose of the tool.

The handle of the gouge is then swung well to the back of the lathe or to the operator's right. The gouge is then pushed forward into the stock and to the left, making a shearing cut. The cut should not be too heavy. The starting point for this cut should be a line which will indicate the largest diameter or circle that can be made from the block.--This cut should be repeated until the corners are removed from the block.

To complete the cutting of thick stock it will be found necessary to change the tool rest to an angle of 45° with the bed of the lathe.

Fig. 14.

When hardwood is being turned it is sometimes advisable to saw the block almost round with a compass saw or bandsaw, if one is to be had. Should this be done the preceding steps are omitted.

The tool rest is then placed parallel with the lathe bed and a roughing cut is taken with the gouge the entire thickness of the block.

The lathe should be run on second or third speed until the corners are removed, and then changed to first speed.

2. CALIPERING FOR DIAMETER. The true diameter is then calipered the same as in spindle work.

3. SMOOTHING CUT. A smoothing cut is taken with a skew chisel the same as in spindle work.

Fig. 15.

4. ROUGHING CUT ON THE FACE. (¾" GOUGE.) FIG. 15. The rest is now placed parallel to the bed of the lathe and slightly above the center of the spindle. Place the gouge on the rest on its edge with the grind toward the stock and parallel to the face to be surfaced. The nose of the gouge is the cutting point.

The handle is then raised and the cutting point is forced toward the center. A very thin shaving should be taken. If the gouge is allowed to roll back so the grind above the cutting point comes in contact with the wood it is sure to catch and gash the wood.

5. SMOOTHING THE FACE. (SMALL SKEW CHISEL.) FIG. 16. For all work up to 3" in diameter, the surface may be smoothed by using a small skew chisel in the same manner as in squaring the ends of Stock in cylinder work. (Step 6--Exercise A-I--1-a, Straight Cuts.)

For larger work, place the chisel flat on the rest with the toe next to the stock and the back edge of the chisel parallel to the face to be surfaced.

The point of the chisel is then forced toward the center of the stock, using the straight back of the tool as a guide against the finished surface. Only a very thin cut should be taken at a time.

Fig. 16.

Note:--While this operation may be termed a scraping cut, it will be found to be much easier on the tool than if the cutting edge were held flat against the work as in other scraping cuts.

The surface of the work should be tested for squareness by holding the edge of the chisel or a straight edge across the face.

LAYING OFF MEASUREMENTS

In laying off measurements on the face of the stock a pencil compass or dividers should be used. Set the compass or dividers to one-half the diameter of the circle wanted. While one point is held at the exact center of the stock, which is easily located while the stock is revolving, the other is brought in contact with the revolving stock until a circle of the correct diameter is marked.

Fig. 17.

Should the center of the stock be cut away, rendering this method impossible, the following method may be used: Set the compass or dividers to the exact diameter wanted. Place one point in contact with the stock a little to one side of the required line on the part that is to be cut-away. Bring the other point to the stock and see if it touches the line first made. If not, move the first point until the two points track in the same line.

Fig. 18.

The rest should be set at the exact center for measuring.

All measurements on the edge of the stock can be made with pencil and rule as in cylinder turning.

B-I--2-a. Shoulder Cuts

1. EXTERNAL SHOULDERS. FIG. 18. The surplus stock at each successive shoulder is roughed out with a ¾" gouge, keeping well outside the finished measurements. The gouge for this work is held in the same position as described in B-I--1-a, Step 1, for Roughing Off Corners.

Fig. 19.

2. For the finishing cut a small skew chisel is used, and the process is the same as that used in squaring ends of stock. Both the vertical and horizontal shoulders can be handled easily by this method. Fig, 19.

3. INTERNAL SHOULDERS. For internal shoulder cutting the same methods may be used for roughing out and cutting the horizontal shoulders, but for the vertical or base shoulder it will be necessary to use the scraping process. (See "Use of Scraping Tools.")

B-I--3-a. Taper Cuts

Taper cutting will not be found hard as the gouge and skew chisel are used in the same manner as described in B-I--1-a, Steps 4 and 5.

After the stock has been roughed away with the gouge to the approximate angle desired, a smoothing cut is taken with the skew. Care should be taken that the skew chisel is held at the exact angle of the taper desired.

B-I--4-a. V Cuts

V cutting will also be found easy as the tool process is exactly the same as that used in spindle turning. Exercise A-I--4-a. Fig. 20.

B-I--5-a. Concave Cuts

Place the ¾" gouge on the rest with the handle parallel to the bed of the lathe. Roll the gouge on its edge and swing the handle so that the grind is perpendicular to the stock with the nose of the tool as the cutting point.

Fig. 20.

Force the gouge forward into the wood. As soon as the cut is started, the handle is lowered and swung to the left; (if cutting the left side of the concave) at the same time the tool is rolled back toward its original position. This movement brings the cutting point farther down on the lip and the grind, resting on the side of the cut, will force the gouge sidewise and will form one-quarter of the circle. Fig. 21.

Fig. 21.

This cut is continued from alternate side until the concave is nearly to size. The cut should be tested with a templet before the finishing cut is taken.

B-I--6-a. Convex Cuts

Rough out the stock between the beads with a parting tool.

Hold the edge of the gouge on the rest with the handle, parallel to the bed of the lathe, to make the nose the cutting point.

Swing the handle to the left so that the grind will form a tangent to the bead at its highest point.

The gouge is then forced into the stock and to the right; at the same time the handle is swung to the right; keeping the grind tangent to the bead at the point of contact. Fig. 22. This cut is continued until the base of the bead is reached.

B-I--7-a. Combination Cuts

As in spindle turning, a combination exercise should be given at this point to provide an opportunity for studying out the best methods of working the various cuts just described into a finished product.

USE OF SCRAPING TOOLS

When scraping is to be employed, it should be done with only those tools that are made for that purpose, i.e., Square Nose, Round Nose, Spear Point, Right and Left Skew. The handling of these tools will be found easy. The only point to remember is that they should be held flat on the tool rest and parallel to the bed of the lathe when in use.

In general practice the ordinary skew chisel should not be used as a scraping tool, for the cutting edge is not sharpened to withstand the heavy strain required by such work. Should it be necessary, however, to use a skew chisel as a scraper, the tool should be held so that the top grind is parallel to the bed of the lathe while in use.

INTERNAL BORING

In roughing out the center for Napkin Rings, Jewel Boxes, etc., the quickest method is to work it out with a small gouge.

Place the gouge on the rest parallel to the bed of the lathe, having the point even with the center of the stock.

Force the gouge into the wood until a hole is bored to the depth required. If the hole is deeper than 1", remove the tool often and clear out the shavings in order not to burn the point.

In order to enlarge the hole to the proper size the point of the gouge is pressed against the left side of the hole a little above the center and a shearing cut is taken. To obviate the danger of the tool catching, all cuts should start from the back of the hole and proceed toward the front.

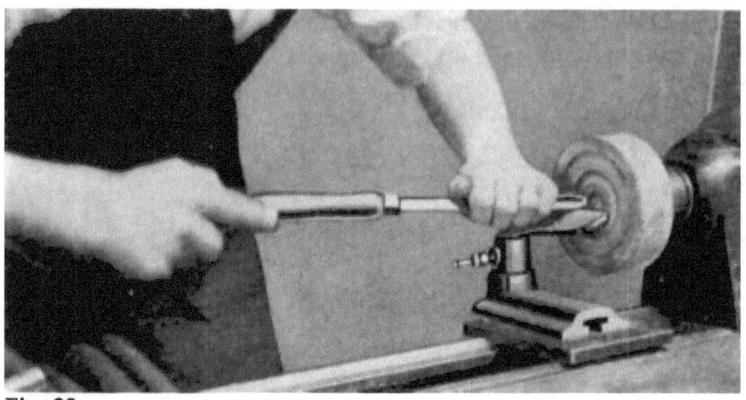

Fig. 22.

B-III--8-a. Sphere

After the sphere is turned as nearly perfect as is possible when working between centers (Steps 1 to 4) it is cut free from the waste stock and is centered in a chuck.

The chuck is made of any soft wood and should be cut in the end grain, which will insure equal pressure on all sides. Equal pressure cannot be obtained if the Chuck is cut in cross grain wood, owing to the tendency of side grain to give more than the end grain. The sphere should be forced into the chuck with slightly over half protruding. Very thin cuts should be taken and the sphere should be revolved one-quarter turn after each until true. As the sphere becomes smaller during the cutting, it will be necessary to cut the face of the chuck down and bore the hole deeper and smaller in order to keep more than half of it protruding at all times.

Mirror (See Pages 299-301).

To remove the sphere tap the chuck lightly with a hammer just above it, at the same time pull out on the sphere.

CHAPTER XI

SPIRAL TURNING

Spiral turning is a subject that has received very little attention by most schools in which wood turning is taught. Spiral work is seen in antique furniture and also in the modern furniture of the present day. It seems that it takes the wheel of fashion about a century to make a complete turn, for what our forefathers neglected and destroyed the people of the present day value and cherish.

Spiral work gives excellent practice in shaping and modelling wood. It brings into play the principle of the helix as used in cutting threads, etc.; and its form, size and shape may be varied according to the taste of the individual. As in threads so in spiral work we have single and double spirals, and their form and proportion depend upon their use and application in furniture making. A variation of the spiral may be made in several ways: First, by changing the number of turns of the spiral on a straight shaft; second, by running a spiral on a tapered shaft; third, by changing the shape or form of the spiral itself; and fourth, by making more than one spiral on a shaft. It is uncommon to see ten or twelve spirals running around a single shaft.

Some of the forms of the above types are fully taken up and explained in the work that is to follow.

PLATES B-V--1-a, B-V--1-a'. SINGLE SPIRAL. STRAIGHT SHAFT

To work out a single spiral for a pedestal proceed as follows:

1. Turn a cylinder 2¼" in diameter. Make the ends slightly larger in order that the design may be turned on each, after the spiral has been worked out.

2. Lay off spaces 2-1/16" apart on the cylinder while the spindle is turning in the lathe and divide each of these into four equal parts. Each one of these large spaces represents one turn of the spiral. A good proportion is slightly less than the diameter of the cylinder; thus the diameter of the cylinder equals 2¼" and the width of the space 2-1/16".

3. On the cylinder parallel to the axis draw lines A-A B-B C-C D-D. These lines should be 90° apart as shown in the top diagram (Plate B-V--1-a'). Line D-D is on the other side of the cylinder as shown in the top and middle diagrams.

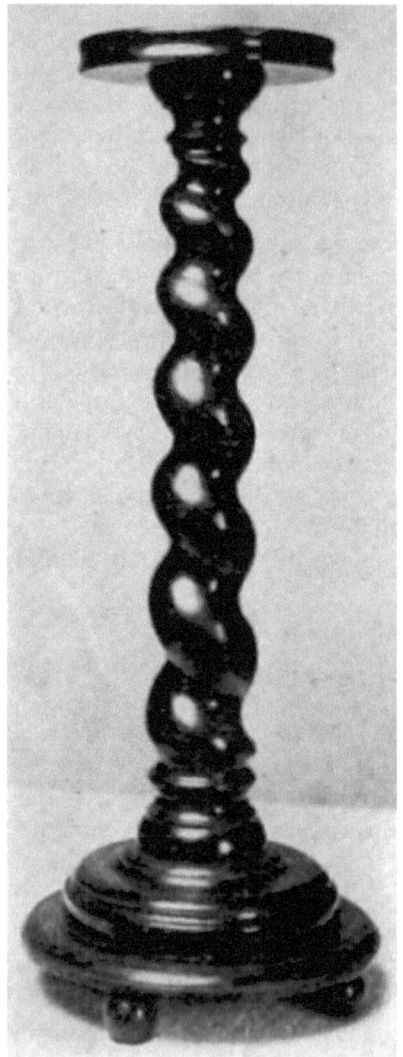

Fig. 23.

4. Start on line A-A at point X, circle 1, and draw a line connecting it with line B-B on circle 1'. Then connect B-B on circle 1' with C-C on circle 2 and so on until a spiral has been drawn the entire length of the cylinder. This line will form the ridge of the spiral as shown in the middle diagram.

5. Next begin on line C-C at circle 1, and draw a line connecting it with D-D on circle 1' then to line A-A on circle 2, and so on as before. This spiral represents the center of the groove or the portion which is to be cut away. This is not shown in the diagram because more or less confusion would be caused with the line representing the ridge of the spiral.

6. Begin on line C-C at circle 1, and saw to a depth of ¾". Saw the entire length of the cylinder leaving about 1½" at the ends. Do not follow the line here, but switch off gradually and follow circles 1 and 15, so as to allow the spiral to begin and end gradually and not abruptly.

7. Rough out with a knife or chisel by cutting on both sides of the saw cut. Then use a wood rasp to finish shaping out the spiral. When properly shaped out allow the lathe to turn slowly and smooth with sandpaper by following the spiral as the lathe turns.

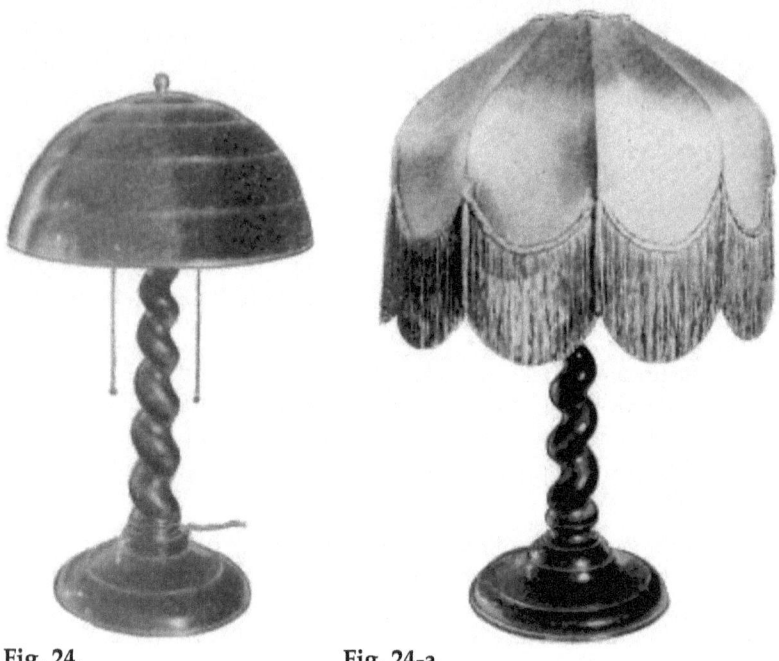

Fig. 24.   **Fig. 24-a.**

8. Cut the design on both ends of the cylinder and polish.

PLATES B-V--2-a, B-V--2-a', B-V--2-a'';. SINGLE SPIRAL. TAPERED SHAFT

To lay off a single spiral for the electric lamp shown in Figs. 24 and 24a proceed as follows:

1. Select your wood and bore a hole through it. Plug the hole and center the piece in the lathe. This insures getting the hole exactly in the center, and it will not be cut into while the cutting of the groove of the spiral proceeds. A groove may also be cut in two pieces of stock and glued together to form a hole through the stock.

2. Turn a cylinder 2½" in diameter, tapering it to 1½" at the one end; this part should be 12⅛" long. Both ends should be left larger than 2½" as the lower and upper designs must be cut here.

3. Let the spindle revolve in the lathe and draw circles as shown in the layout (Plate B-V--2-a'). The number of circles will vary with the taper. Since seven turns are needed in the present spiral, 28 circles will be necessary--four circles for each turn of the spiral as shown in the middle diagram. A good proportion to follow is to measure the diameter of the spindle at circle 2 and lay off this distance from circle 1 to circle 3. Then measure the diameter at circle 4 and lay off this diameter from circle 3 to circle 5 and so on until all circles have been made. Then divide these large divisions into four equal parts.

4. Draw four lines the entire length of the spindle, each 90° apart as shown by the heavy lines in the middle diagram. The heavy circles of the same diagram represent the complete turns of the spiral.

5. Lay out the line representing the ridge of the spiral as shown in the middle diagram. Begin on circle 1, where the straight line crosses it, draw to circle 1' at the point where the next straight line crosses it, then to 2--2'--3--3' and so on until the end is reached. This forms the ridge of the spiral as shown in diagram 3. Next it may be more convenient to draw another line representing the groove. In this case begin at point X in the middle diagram, opposite the point where first started, and continue in the preceding manner, making this line parallel to the other line.

6. Saw on the line last made, being careful not to saw too deeply. The depth must be ¼" less than half the diameter of the spindle where the cut is made. This saw cut forms the groove of the spiral. The groove is then cut out by hand with a chisel or knife, by working down the wood on both sides of the saw cut. After the spirals have been roughed out, a rasp is used to finish shaping them. The work is then sandpapered smooth, while the spindle is revolved slowly in the lathe.

7. Cut designs on the ends of the cylinder and polish.

PLATES B-V--2-b, B-V--2-b'. DOUBLE SPIRAL. TAPERED SHAFT

To work out a double spiral for the electric lamp illustrated in Fig. 25 proceed as follows:

1. Turn up the spindle in the usual manner. Since the base of the shaft is larger than the top, the spiral must also be in proportion and lines A-A', B-B', C-C', D-D', and E-E', are drawn around the shaft. To get the approximate spacing from circles A-A to B-B measure the diameter at A-A' plus about 3/16" and lay off from A-A' to B-B'. Then take the diameter of B-B' plus about 3/16" and lay off from A-A' to B-B'. Then take the diameter at B-B' plus about 3/16" and lay off from circle B-B' to C-C' and so on. If the shaft is tapered more, a different proportion must be used. Also if it is desired to have the twist wind around the shaft three times, a variation must be made in the number of circles.

Fig. 27.

2. If it is desired to have the twist wind around the shaft twice, draw circles 1-1', 2-2', 3-3', and 4-4' and the spaces will grow proportionately smaller at the small end.

3. Draw four lines running lengthwise on the spindle and 90° apart as shown in the midde figure in heavy lines (Plate B-V--2-b').

4. Begin at A and draw a curved line to where the 90° line crosses circle 1-1'. From there extend the line to where the next 90° line crosses circle B-B' at point B'. Continue in this manner until the other end of the shaft is reached. Begin at A' and draw a line on the opposite side of the shaft. These two lines running around and along the shaft form the grooves while the portion in between forms the beads of the double spiral.

5. Saw to the desired depth, being ¼" less than half the diameter at the point where cut. With a chisel or knife form the grooves and beads. It is necessary to be careful about not ending the grooves too abruptly. (See point 6 in Plates B-V--1-a, B-V--1-a'.) Smooth with a rasp and sandpaper while the lathe is revolving slowly.

6. Cut the design on the ends and polish.

PLATES B-V--3-a, B-V--3-a'. DOUBLE GROOVE SPIRAL. STRAIGHT SHAFT

To work out the double groove spiral for the magazine holder illustrated, proceed as follows:

1. Square up the stock to 1⅜". Center carefully and turn the design on both ends as shown, in the upper diagram (Plate B-V--3-a'). Turn the cylinder between the top and bottom, making it 5½" long and 1⅜" in diameter.

Fig. 26.

2. Divide the cylinder into two equal parts. Each part represents one revolution of the spiral.

3. Divide each half into four equal parts as shown in the top and center diagrams (Plate B-V--3-a'), 1-1', 2-2', 3-3' and so on. The proportion of the distance between these circles should be one-half the diameter of the cylinder.

4. Draw lines A-A, B-B, C-C, and D-D, parallel to the axis of the cylinder 90° apart.

5. With a band 3/16" wide of any substantial material (preferably a narrow strip of tin or a watch main spring) begin on the line A-A at circle 1, and connect circle 1' at line B-B, and then connect circle 2

at C-C, and so on until the spiral is made the entire length. Mark on both sides of the 3/16" band so as to keep the spiral parallel.

6. Next begin at the line C-C where circle 1 crosses it and connect from here to 1' at B-B. Proceed as in Step 5, as shown in the center diagram.

7. Now erase the extreme ends of the spiral near circles 1 and 5, and deviate from the original spiral and follow the circles in a more parallel direction so as to allow the spiral to begin and end gradually and not too abruptly. Refer to the lower diagram for this.

8. Cut out portions of wood between the bands previously marked around, as shown in the lower figure. The wood should be cut out with a knife so as to leave the corners sharp on the narrow bands. The portion cut out should be a semi-circle and can be sanded by making a spindle a little smaller than the distance between the bands and fastening sandpaper on the spindle. Place in the lathe and hold the spiral on the sandpaper cylinder at an angle so that the spiral will fit. Turn gradually and the sandpaper will smooth up the portion between the bands and true it up. At the ends where the grooves are smaller, use a smaller stick around which sandpaper has been wound and work out by hand.

9. It is well to cut straight down, about 1/32" deep, along the lines marking out the narrow bands. Then the wood will not be so likely to split while removing the stock which forms the grooves between the bands.

10. Cut out the mortises in the square portions which have been left at both ends. Make the frame work for the sides and cane. Glue together and polish.

Note:--By making the posts smaller and using the same construction for a side a nice looking book stall may be made. The proportions for the posts are the same as mentioned in Step 3.

[Transcribers note: There are 142 line art illustrations after this point in the book. See the Classification of Plates for all of them.]

www.ingramcontent.com/pod-product-compliance
Lightning Source LLC
Chambersburg PA
CBHW030451220526
45464CB00006B/2493